Brick M...

LEARNING DIVISION

USING LEGO® BRICKS

STUDENT EDITION

Dr. Shirley Disseler

COMPASS

Learning Division Using LEGO® Bricks — Student Edition

Brigantine Media/Compass Publishing
211 North Avenue
St. Johnsbury, Vermont 05819
Phone: 802-751-8802
Fax: 802-751-8804
E-mail: neil@brigantinemedia.com
Website: www.compasspublishing.org

ORDERING INFORMATION
Quantity sales
Special discounts for schools are available for quantity purchases of physical books and digital downloads. For information, contact Brigantine Media at the address shown above or visit www.compasspublishing.org.

Individual sales
Brigantine Media/Compass Publishing publications are available through most booksellers. They can also be ordered directly from the publisher.
Phone: 802-751-8802 | Fax: 802-751-8804
www.compasspublishing.org
ISBN 978-1-9384066-0-7

CONTENTS

DIVISION FACTS

1. Place a 2x4 brick on a base plate. Draw your model.

How many studs are on the brick? _____

Locate four same-sized bricks that you can place on top of the 2x4 brick with no stud left uncovered. Place your bricks on top of the 2x4 brick. Which bricks did you use? _____ How many bricks did you use? _____

Write a division sentence for this problem. _____

Explain your thinking.

_____

_____

_____

2. Place a 1x6 brick on a base plate. Draw your model.

Find three same-sized bricks that fit on top of the 1x6 brick with no stud left uncovered. Place them on top of the 1x6 brick. Which bricks did you use? _____ How many bricks did you use? _____

Write a division sentence for this problem. _____

Explain your thinking.

_____
_____
_____

Place 1x3 bricks on top of the 1x6 brick to cover the brick. How many bricks did you use? Write the division sentence. _____
Draw your model.

Place another set of bricks that covers the 1x6 brick. Build a model using those bricks. Which bricks did you use? _____ Draw your model.

Write all the division sentences for the 1x6 model.

_____

_____

_____

3. Place a 2x4 brick on a base plate. What is the whole represented by this brick? _____

Find two same-sized bricks that fit on top of this brick with no stud left uncovered. Place them on top of the 2x4 brick. Which brick did you use? _____
How many bricks did you use? _____

Write a division sentence for this problem. _____

Draw your model and explain your thinking.

Look back at problem 1. How is this problem different?

_____

_____

_____

Find other bricks that work for the 2x4 brick. Build a model and draw your solution.
Which bricks did you use? _____

List all the solutions for the 2x4 brick.

_____

_____

_____

4. Place a 2x3 brick on the base plate. What is the whole represented by this brick? _____

Find three same-sized bricks that will fit on top of this brick with no stud left uncovered.
Place them on top of the 2x3 brick. Which brick did you use? _____ How many bricks did
you use? _____

Write a division sentence for this problem. _____

Draw your model and explain your thinking.

Find other bricks that cover the 2x3 brick. Which bricks did you use?_____

Draw your model. Explain your thinking. List all the division sentences for the 2x3 brick.

More problems for practice:

Using the process from problems 1 – 4, build models for the following bricks:

5. 1x16

Total number of studs representing the whole: ___________

With bricks, model how to divide this brick using 1x2 bricks.
How many bricks did you use? ______

Draw your model and explain your thinking.

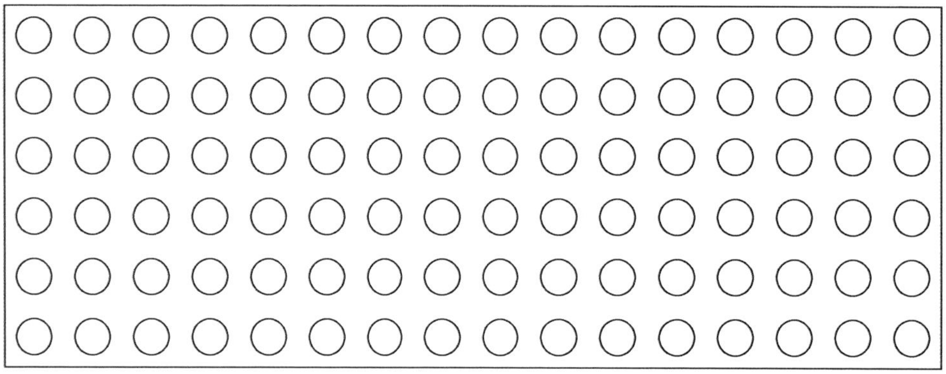

____

____

____

Write a division sentence for your model. ____

6. 2x8

Total number of studs representing the whole: ___________

With bricks, model how to divide this brick using 2x2 bricks. How many bricks did you use? ______

Draw your model and explain your thinking.

_____
_____
_____
_____
_____
_____
_____
_____
_____
_____
_____
_____
_____
_____

Write a division sentence for your model. _____

Assessment:

1. What does it mean to divide?

_____
_____
_____

2. Use a 2x8 brick to show a division sentence with a solution of 8. Draw your model. Which bricks did you use and what do they represent?

_____
_____
_____
_____
_____
_____
_____
_____
_____
_____
_____

FINDING FACTORS

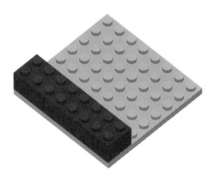

1. With bricks, build this model.

How many studs are on the brick? _____ What does the number of studs mean? _____

_____

Locate two same-sized bricks that together are the same length and width as the 2x8 brick. Place them next to the 2x8 brick. Which bricks did you use? _____ How many bricks did you use? _____ What do these two bricks mean?

_____

_____

_____

Write a division sentence for this model. _____

Explain your thinking.

_____

_____

_____

Can you find three same-sized bricks that will fit into the model? _____

Find four same-sized bricks that will fit next to the model that together are the same length and width as the last bricks you added. Place them on your model next to the bricks. Which bricks did you use? _____   How many bricks did you use? _____

Write a division sentence for this problem. _____

Explain your thinking.

_____

_____

_____

What other bricks work with this model? _____
Add those bricks to your model.

Write division sentences for these brick models. What do you notice about all the bricks on the base plate model?

_____

_____

_____

Write all the division sentences for the 2x8 model.

_____

_____

_____

What do all of these statements have in common? _____

What are the factors of 16? _____

2. With bricks, build a model to show all of the fact families/factors of 6. Draw your model and explain your thinking.

_____
_____
_____
_____
_____
_____
_____
_____

Write all of the division sentences for this problem.

_____
_____
_____

What are the factors of 6? _____

3. With bricks, build a model to show all the ways to divide 8. Draw your model and explain your thinking.

_____
_____
_____
_____
_____
_____
_____
_____

Write all the division sentences for this problem.

_____

_____

_____

What are the factors of 8? _____

More problems for practice:

4. With bricks, build a model to show all the ways to divide 12. Draw your model.

Write all the division sentences for the model.

_____

_____

_____

Write all the factors of 12. _____

5. With bricks, build a model to show all the ways to divide 4.

○ ○ ○ ○ ○ ○ ○ ○
○ ○ ○ ○ ○ ○ ○ ○
○ ○ ○ ○ ○ ○ ○ ○
○ ○ ○ ○ ○ ○ ○ ○
○ ○ ○ ○ ○ ○ ○ ○
○ ○ ○ ○ ○ ○ ○ ○
○ ○ ○ ○ ○ ○ ○ ○
○ ○ ○ ○ ○ ○ ○ ○

Write all the division sentences for the model.

_____

_____

_____

Write all the factors of 4. _____

Assessment:
1. What is a factor?

_____

_____

_____

2. How are factors used in division?

_____

_____

_____

3. With bricks, build a model to show all the factors of 24. (Note: you will need to use more than one brick in your whole. Make the whole one color to make this easier to see.) Draw your model.

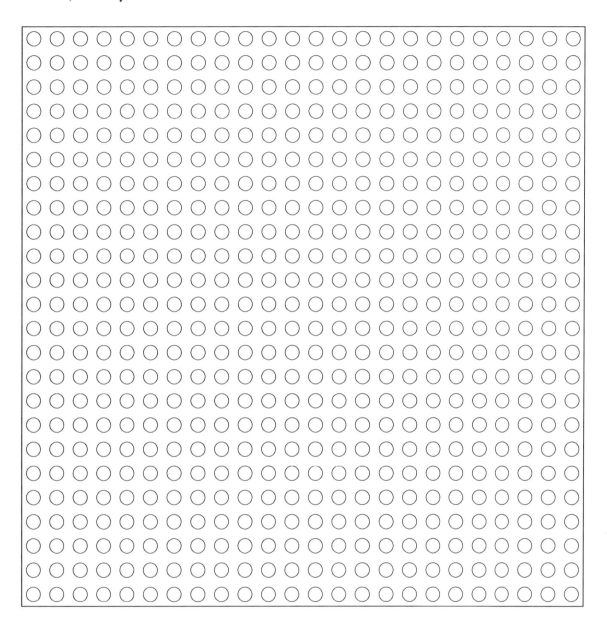

Explain the parts of your model and what they represent. Be sure to list all the facts.

_____

_____

_____

3

EXPLORING DIVISION

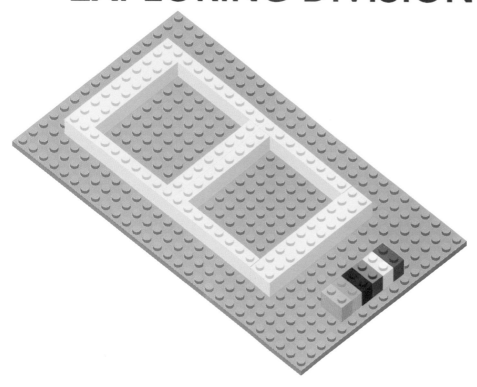

1. With bricks, build this model. How many total studs are on the 1x2 bricks? _____
What do the number of studs mean on these bricks? _____
_____

2. Distribute the 1x2 bricks evenly into the two set boxes. How many studs are in each set box? _____

Write a division sentence for this model. _____

Explain your thinking.

_____
_____
_____

Write a multiplication sentence for this problem. _____

How are these two sentences related to one another?

_____

_____

_____

The set boxes represent what division word? _____

The studs inside the boxes represent what division word? _____

3. Place six 1x2 bricks in a row on a base plate. Take away bricks, one at a time. Complete the chart:

Number of studs	Number removed (subtracted)	Number Left
I started with _____ studs	I took away 2 studs	Now I have _____ studs
I have ____ studs	I took away 2 studs	Now I have _____ studs
I have ____ studs	I took away 2 studs	Now I have _____ studs
I have ____ studs	I took away 2 studs	Now I have _____ studs
I have ____ studs	I took away 2 studs	Now I have _____ studs
I have ____ studs	I took away 2 studs	Now I have _____ studs

How many times did you subtract 2 studs from the original model? _____

How is this related to dividing 12 by 2?

_____

_____

_____

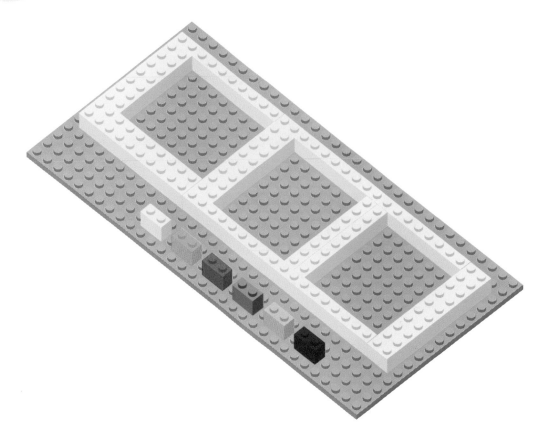

4. With bricks, build this model. Place an even number of 1x2 bricks in each set box. How many bricks did you place in each set box? _____

How many studs are in each set box? _____

Write a division sentence for this problem. _____

Explain your thinking.

_____

_____

_____

Write a multiplication sentence for this model. _____

Complete the table to show repeated subtraction as a solution for this problem:

Number of studs	Number of studs removed (subtracted)	Number of studs left
I have 12 studs	I took away _____ studs	Now I have _____ studs
I have _____ studs	I took away _____ studs	Now I have _____ studs
I have _____ studs	I took away _____ studs	Now I have _____ studs

Challenge: Create another way to divide 12 using the same 1x2 bricks. You can add set boxes to your model. Draw your model and explain your thinking.

5. With bricks, build a model that shows 24 divided by 2. Build the set boxes that show the divisor of 2. Choose bricks of the same size to show 24 studs. Place them into the set box model to indicate division into sets.

Draw your model and explain your thinking.

_____
_____
_____
_____
_____
_____
_____
_____
_____
_____

Write a division sentence and a multiplication sentence for this problem.

_____   _____

Draw a table to show repeated subtraction for your model.

Number of studs	Number removed (subtracted)	Number left

6. With bricks, build a model that shows 15 divided by 3. Build the set boxes that show the divisor of 3. Choose bricks of the same size to show 15 studs. Place them into the set box model to indicate division into sets.

Draw your model and explain your thinking.

Write a division sentence and a multiplication sentence for this problem.

_____   _____

Draw a table to show repeated subtraction for your model.

Number of studs	Number removed (subtracted)	Number left

7. With bricks, build a model that shows 18 divided by 3. Build the set boxes that show the divisor of 3. Choose bricks of the same size to show 18 studs. Place them into the set box model to indicate division into sets.

Draw your model and explain your thinking.

_____

_____

_____

_____

_____

_____

_____

_____

_____

_____

_____

_____

_____

Write a division sentence and a multiplication sentence for this problem.

_____   _____

Draw a table to show repeated subtraction for your model.

Number of studs	Number removed (subtracted)	Number left

More problems for practice:

8. With bricks, build a model that shows 16 divided by 4. Build the set boxes that show the divisor of 4. Choose bricks of the same size to show 16 studs. Place them into the set box model to indicate division into sets.

Draw your model and explain your thinking.

_____

_____

_____

Write a division sentence and a multiplication sentence for this problem.

_____   _____

Draw a table to show repeated addition for your model.

Number of studs	Number removed (subtracted)	Number left

Assessment:

1. Circle the divisor in each problem.

$12 \div 4 = 3$
$20 \div 5 = 4$

2. Circle the dividend in each problem.

$12 \div 3 = 4$
$20 \div 5 = 4$

3. With bricks, build a model to show $8 \div 2$. Draw your model and explain your thinking.

Write a multiplication sentence for the problem. _____

4. Draw a table to show the repeated subtraction for $8 \div 2$.

Number of studs	Number removed (subtracted)	Number left

5. Explain how repeated subtraction and division are related.

_____

_____

_____

6. How are multiplication and division related?

_____

_____

_____

EQUAL SHARES OR PARTITIVE DIVISION

Partitive Division: Sharing equally among sets

1. With bricks, build this model. It represents 12 pieces of candy (each stud is one piece). Find two same-sized bricks that have the same total number of studs as the one in your model. Place these two on the model next to the 12-stud brick. Each of these bricks represents one friend who will get an equal share of the 12 pieces of candy.

How many pieces will each friend get? _____
Draw your model. Using your model, explain your thinking.

_____

_____

_____

_____

_____

_____

2. Now you have 4 friends to share the 12 pieces with equally. With bricks, build a model to show how to equally share the 12 pieces of candy with your 4 friends. Draw your model and explain your thinking.

3. Place a 2x3 brick on a base plate and show how to divide this amount equally among 2 friends, 3 friends, and 6 friends.

Draw each model and explain your thinking.

_____

_____

_____

Another term for this process of equal sharing is called _____ division.

4. Build a model with two 2x6 bricks to show a whole of 24.

Show four ways to use partitive division to equally share 24. (Note: There are more than four ways.)

Draw your models and explain your thinking.

_____
_____
_____
_____
_____
_____

5. Build a model with one 2x4 brick to show a whole of 8. Show two ways to equally share this whole by partitioning it into equal parts on base plates. Draw your models and explain your thinking.

_____

_____

_____

6. With bricks, build a model that shows the whole as 16. Partition the whole into equal shares in at least two ways. Draw your models and explain your thinking.

_____

_____

_____

7. Choose a number greater than 24. Using bricks, model the number on a base plate. Show at least two ways to equally share this amount. Draw your models and explain your thinking.

More problems for practice:

8. With bricks, build a model that shows a whole of 10. Show two ways to partition this whole into equal shares. Draw your models and explain your thinking.

9. Choose any number that can be shared equally. With bricks, build a model to show the whole. Show two ways to partition this whole into equal shares. Draw your models and explain your thinking.

Assessment:

1. What does it mean to share something equally?

_____

_____

_____

2. What does the term *partition* mean?

_____

_____

_____

3. With bricks, build a model that shows 4 candy bars equally shared with 2 friends. Draw your model and explain your thinking.

_____

_____

_____

_____

_____

_____

_____

_____

_____

_____

4. With bricks, build a model to show 9 shared equally among 3 friends. Draw your model and explain your thinking.

_____

_____

_____

_____

_____

_____

_____

_____

_____

REPEATED SUBTRACTION OR QUOTITIVE DIVISION

Quotitive Division: dividing by repeating subtraction

1. With bricks, build this model. How many studs are on this brick? _____

If each stud represents pieces of candy, how many pieces of candy do you have? _____

If you want to give your friends 4 pieces of candy each, how many friends can you share your candy with? _____

What is the repeated subtraction sentence for this model? _____

Write a division sentence for this model. _____

Use repeated subtraction to model this process with bricks.

Draw your model and explain your thinking.

_____
_____
_____
_____
_____
_____
_____
_____
_____
_____
_____
_____
_____
_____
_____

2. Place a 2x8 brick on a base plate. What whole does the brick represent? _____

If each stud represents one dollar, and you want to purchase tickets for carnival rides that cost $2 each, how many tickets can you buy? _____

With bricks, model the problem using the process of repeated subtraction. Draw your model and explain your thinking.

○○○○○○○○○○○○○○○○○○ _____
○○○○○○○○○○○○○○○○○○ _____
○○○○○○○○○○○○○○○○○○ _____
○○○○○○○○○○○○○○○○○○ _____
○○○○○○○○○○○○○○○○○○ _____
○○○○○○○○○○○○○○○○○○ _____
○○○○○○○○○○○○○○○○○○ _____
○○○○○○○○○○○○○○○○○○ _____
○○○○○○○○○○○○○○○○○○ _____
○○○○○○○○○○○○○○○○○○ _____
○○○○○○○○○○○○○○○○○○ _____
○○○○○○○○○○○○○○○○○○ _____
○○○○○○○○○○○○○○○○○○ _____
○○○○○○○○○○○○○○○○○○ _____
○○○○○○○○○○○○○○○○○○ _____
○○○○○○○○○○○○○○○○○○ _____

Write a repeated subtraction sentence for this model. _____

Write a division sentence for this problem. _____

3. Place a 2x4 brick on a base plate. What whole is represented by this brick? _____

How many equal groups of 4 are in this whole? _____

With bricks, build a model to show your solution using the repeated subtraction process. Draw your model and explain your thinking.

Write a repeated subtraction sentence for this model. _____

Write a division sentence for this problem. _____

More problems for practice:

4. Place a 2x6 brick on a base plate. What whole is represented by this brick? _____

If each stud is equivalent to one dollar, how much money do you have? _____

With bricks, build a model using repeated subtraction to show how many movie tickets you could buy if each ticket cost 4 dollars. Draw your model and explain your thinking.

5. Choose a combination of bricks that shows a whole of 18. With bricks, build models using repeated subtraction in two different ways. Draw your models. Write a repeated subtraction sentence and a division sentence for each model. Explain your thinking.

Assessment:

1. Explain how division is like repeated subtraction.

_____

_____

_____

2. If you have 8 feet of rope and you cut it into 2-foot pieces, how many pieces will you have? With bricks, build a model and show your solution using repeated subtraction. Draw your model and explain your thinking.

_____

_____

_____

_____

_____

_____

_____

_____

_____

3. Write a division sentence for the repeated subtraction problem

$24 - 3 - 3 - 3 - 3 - 3 - 3 - 3 - 3 = 0$ _____

4. With bricks, model problem 3 using repeated subtraction. Draw your model and explain your thinking.

_____

_____

_____

_____

_____

_____

_____

_____

_____

TWO-DIGIT DIVISION

1. Model the number 24 as the whole using two 2x6 bricks. Since 24 is the number that will be divided into parts, it is called the _____.

Place 1x6 bricks on top of the 24 studs to cover all the studs. Each 1x6 brick represents the number _____ because of the number of studs on them. This number is called the _____. How many 1x6 bricks did you place on your model? _____ This number is called the _____.

Write a division sentence for this model. _____

Draw your model.

2. With bricks, build a model that has 48 studs. This is the dividend. Using a divisor of 8, how many times can you cover this model? _____ Draw your model and explain your thinking.

_____
_____
_____
_____
_____
_____
_____
_____

Write a division sentence for this model. _____

3. Use the dividend of 48 again, but this time, use a divisor of 4. Model the problem and solution with bricks. How many times does 4 divide into 48? _____ Draw your model and explain your thinking.

_____
_____
_____
_____
_____
_____
_____
_____

4. Model the dividend of 18. Using the divisor of 4, model your solution to the problem. What do you notice?

_____

_____

_____

Draw your model and explain your solution.

_____

_____

_____

_____

_____

_____

_____

What does it mean to have a remainder? _____

More problems for practice:

5. With bricks, model 36 divided by 4. Draw your model and explain your solution. Write a division sentence for your model.

_____

_____

_____

6. With bricks, model 70 divided by 4. Draw your model and explain your solution.

_____
_____
_____
_____
_____
_____
_____
_____
_____
_____
_____
_____

7. With bricks, model the division problem 25 ÷ 2 = _____

Draw your model and explain your thinking.

_____
_____
_____
_____
_____
_____
_____
_____
_____
_____
_____
_____

8. Choose any two-digit number to model. With bricks, model two different division problems for this number that do not have remainders. Draw your models and explain your thinking.

9. Choose a two-digit number to model. With bricks, model the division process using a divisor that provides a remainder. Draw your model and explain your thinking.

_____

_____

_____

10. With bricks, model the number 10. Model a way to divide this number without any remainder. Draw your model on the left diagram. Write the division statement below it.

Use the same number (10) and build a model with bricks of a division problem that has a remainder. Draw your model on the right diagram. Write the division statement below it.

_____   _____

Assessment:

1. Identify each part of this division problem. 40 ÷ 3 = 13 R 1

       Quotient_____

       Divisor _____

       Remainder _____

       Dividend _____

2. What is a remainder and what does it mean?

_____

_____

_____

3. With bricks, model a dividend of 16 and a divisor of 2. What is the quotient? _____
Draw your model and explain your thinking.

_____

_____

_____

7

DIVIDING LARGER NUMBERS

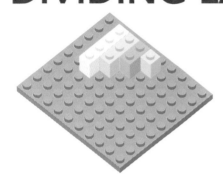

1. With bricks, make a place value model: place a 1x1 brick on the right side, a 1x2 brick to its left, a 1x3 brick to the left of the 1x2 brick, and a 1x4 brick to the left of the 1x3 brick.

The 1x1 brick represents the ones place, the 1x2 brick represents the tens place, the 1x3 brick represents the hundreds place, and the 1x4 brick represents the thousands place.

Draw your model and label the place value of each brick.

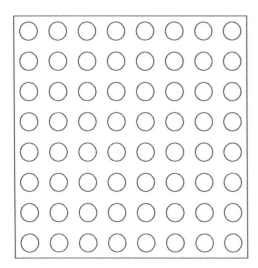

2. With bricks, model the number 24 using the place value method. Explain your model.

_____

_____

_____

_____

_____

_____

_____

_____

_____

3. With bricks, model the number 2,345 using the place value method. Refer to your place value model in problem 1 for guidance.

4. With bricks, model the number 222 using bricks the place value method. The divisor for this problem is 2. Write a division statement for the problem. _____

With bricks, model the division of the hundreds, tens, and ones into 2 sets. Draw your model.

Which bricks are in each set?

_____

What number does each set represent?

_____

What is the quotient?

_____

5. With bricks, model the number 1,031. Divide each place value into two sets to represent dividing by 2. Which bricks are in each set? _____

What do you notice? _____

What number/quotient do the bricks in each set represent? _____

What is the remainder? _____

Write a division sentence for this model. _____

Can you write the remainder as a fraction? _____

Can you write the remainder as a decimal? _____

6. With bricks, model 444 ÷ 4. Show the place value sets. Determine the quotient. Draw your solution model and explain the quotient.

_____

_____

_____

_____

_____

_____

_____

_____

_____

7. With bricks, model 1,246 ÷ 2. Show the place value sets. Determine the quotient. Draw your solution model and explain the quotient.

_____

_____

_____

_____

_____

_____

_____

_____

8. With bricks, model 1,345 ÷ 2. Show the place value sets. Determine the whole number quotient and the remainder. You can define the remainder as a fraction, whole number, or decimal. Draw your solution model and explain the quotient.

_____

_____

_____

_____

_____

_____

_____

_____

More problems for practice:

9. With bricks, model 2,345 ÷ 3. Show the place value sets. Determine the whole number quotient and the remainder. You can define the remainder as a fraction, whole number, or decimal. Draw your solution model and explain the quotient.

_____

_____

_____

_____

_____

_____

_____

_____

_____

_____

10. With bricks, model 2,226 ÷ 4. Show the place value sets. Determine the whole number quotient and the remainder. You can define the remainder as a fraction, whole number, or decimal. Draw your model and explain the quotient.

_____
_____
_____
_____
_____
_____
_____
_____
_____
_____
_____
_____
_____
_____

Assessment:

1. With bricks, model 1,224 using the place value method. Draw your model.

2. With bricks, model 222 ÷ 2 using the place value method. Draw your model and explain your quotient.

_____
_____
_____
_____
_____
_____
_____
_____
_____
_____
_____

3. With bricks, model 1,242 ÷ 4. On your model, show the quotient and the remainder. Draw your model and explain your solution.

_____
_____
_____
_____
_____
_____
_____
_____
_____
_____
_____
_____
_____

Printed in Great Britain
by Amazon